Imagine this...

Stories inspired by Agriculture
2020

Ellie Stover

Valerie Nava

Olivia Piazza

Finley Brady

Isabella Diep

Nathan Tanega

Ellie Gomes

Learn About Ag
California Foundation for
Agriculture in the Classroom

Simplot

Publisher's Cataloging-in-Publication Data

Stover, Ellie
   Imagine this: stories inspired by agriculture 2020 / Ellie Stover, Valerie Nava, Olivia Piazza, Finley Brady, Isabella Diep, Nathan Tanega, Ellie Gomes
   p.cm.
   ISBN 978-0-578-45407-8

   Summary: 2020 winning stories from the California Foundation for Agriculture in the Classroom, Imagine this... Story Writing Contest written by third through eighth grade students.

[1. California–Social life and customs–Fiction. 2. Farm life–California–Fiction. 3. Food–Fiction. 4. Short stories. American.] I. Stover, Ellie. II. Nava, Valerie. III. Piazza, Olivia. IV. Brady, Finley. V. Diep, Isabella. VI. Tanega, Nathan. VII. Gomes, Ellie. VIII. Title.

California Foundation for
Agriculture in the Classroom
2600 River Plaza Drive, Suite 220
Sacramento, CA 95833
(916) 561-5625 • Fax (916) 561-5697
(800) 700-AITC (2482)

LearnAboutAg.org

This book is dedicated to:

The J.R. Simplot Company Foundation
for their on-going support of the
Imagine this… Story Writing Contest.

TABLE OF CONTENTS

Introduction 1

State Winning Stories

The Mystery of Merino's Wool
by Ellie Stover 3

The Strawberry's Owner
by Valerie Nava 17

One Amazing Family
by Olivia Piazza 31

My Journey: A Drop of Water
by Finley Brady 45

The Pistachio Man
by Isabella Diep 59

Laura & Lana: California Fairies of Flora and Fauna
by Nathan Tanega 73

Honorable Mention

Alissa Alfalfa's Big Journey
by Ellie Gomes 86

Highlights

Glossary 90
Acknowledgments 92
About California Foundation for Agriculture in the Classroom ... 93
Imagine this… Story Writing Contest 94
Entry Form 95

INTRODUCTION

The world of agriculture is brought to life through the creative words of talented California students in grades three to eight. The 2019-2020 *Imagine this…* Stories Inspired by Agriculture book showcases seven inspiring stories ranging in topics from alfalfa to wool and even a poem about Flora and Fauna! The stories are illustrated by students in Northern California high school art programs.

More than a thousand students throughout California have annually participated in the statewide contest coordinated by California Foundation for Agriculture in the Classroom. Students select an agricultural topic, research it, and write a creative story to share with others. This contest gives students an opportunity to develop a better understanding about where their food and fiber come from and how they are produced.

The *Imagine this…* Story Writing Contest challenges students to use creative thinking and their imagination to share what they learn through writing. We hope you enjoy the stories, and we encourage you to share your appreciation of agriculture with others.

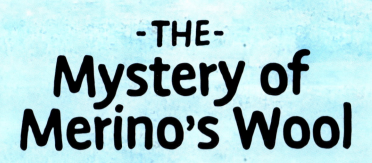

-THE-
Mystery of
Merino's Wool

By Ellie Stover
- 3rd grade -
Plaza School

Jennifer Limberg, Teacher
Glenn County

Illustrated by Inderkum High School

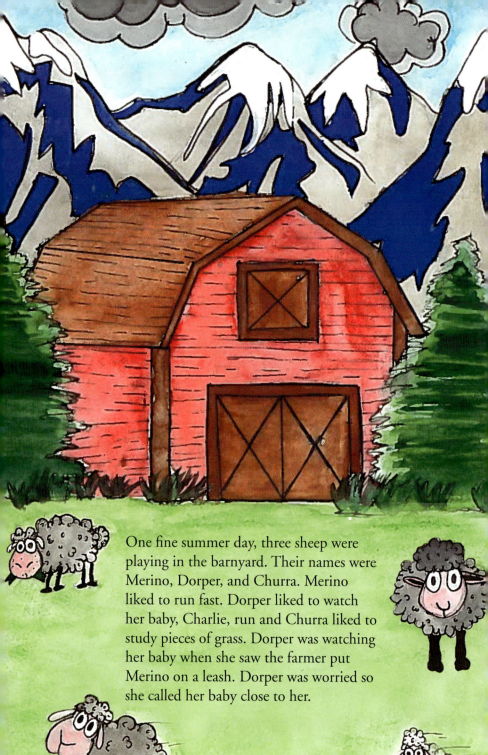

One fine summer day, three sheep were playing in the barnyard. Their names were Merino, Dorper, and Churra. Merino liked to run fast. Dorper liked to watch her baby, Charlie, run and Churra liked to study pieces of grass. Dorper was watching her baby when she saw the farmer put Merino on a leash. Dorper was worried so she called her baby close to her.

4

The farmer took Merino, and they walked down to the barn. When it was time for lunch, Merino came scrambling out of the barn. When Merino came to a stop, he yelled to his friends,

"Hey, Dorper and Churra!"

They looked at him closely before Dorper asked, "Who are you?"

"Me?" replied Merino. "I am your friend, Merino. You know me!"

"We do?" asked Dorper and Churra, confusedly.

"What happened to you?" asked Dorper.

5

"Nothing," said Merino.

"Oh, something happened to you," stated Churra.

Merino looked down at himself and said, "OMG, I'm naked!"

"There are over one billion sheep in the world, and you are not the only one to get sheared," said Churra.

"Sheared! I didn't want to be sheared!" cried Merino.

"Hey, this is a mystery to find Merino's wool the farmer sheared off of him!" yelled Charlie, Dorper's baby lamb. Merino and Churra nodded excitedly.

Merino shouted, "I know where to look first!"

"Where?" asked Dorper and Churra.

"The barn! That's where I went first," said Merino.

"Then let's go! What are we waiting for?" Dorper shouted as she was racing toward the barn.

The farmer saw them running and called, "Get into the barnyard!"

Dorper, Merino and Churra had to listen to the farmer, so they turned around and went back to the barnyard.

"We need a way into the barn," stuttered Charlie frantically. "Yeah, I think I know a way in," Churra wailed.

8

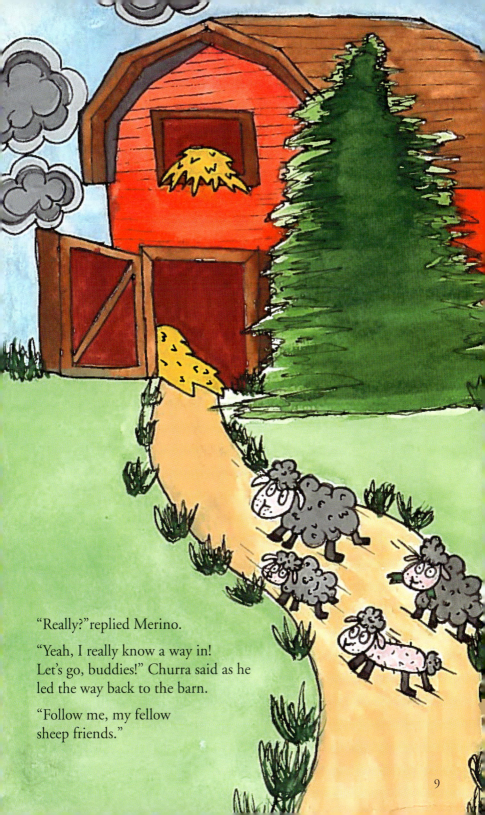

"Really?" replied Merino.

"Yeah, I really know a way in!
Let's go, buddies!" Churra said as he
led the way back to the barn.

"Follow me, my fellow
sheep friends."

9

When they made it to the barn, they slipped in quietly and saw something magical. The reason the farmer sheared Merino was because the farmer's wife wanted to make something special from it. As they watched, the farmer came in with some tea for his wife. He was surprised to see the three sheep. "You better get into the barnyard right now," he said.

The sheep didn't get out of the barn, so the farmer chased the three sheep around the barn!

His wife calmly said, "Let them stay, James."

"Okay, Lily," said James quietly back to his wife.

Lily offered the sheep a handful of hay to bring them closer to her and the baby. Lily showed them the adorable little baby girl that was swaddled in a blanket that Lily had crocheted using Merino's wool. She whispered into Merino's ear softly, "Your wool is the finest and softest wool we have ever been lucky enough to have on our farm. Our baby is so happy and warm. Thank you!"

From then on, the farmer sheared Merino every spring so the new baby girl could have a sweatshirt, mittens, and blanket for every winter. Merino was very happy he could help his beautiful family.

About the Author:

Ellie Stover, age 8

Third grader Ellie Stover decided to write her story, *The Mystery of Merino's Wool*, because she loves animals. She specifically chose sheep and wool because she loves how their wool is very fluffy, soft, and useful. Ellie's teacher gave her examples of past regional winning stories. That helped her get started on her own story. She chose the name Merino because it was interesting that it was the real name of a sheep breed. Her aunt has sheep, and she thought about them while writing her story. Ellie is already thinking about her next story about her aunt's llama that had a baby recently.

Ellie likes school and enjoys writing stories. She planned her story all in her head, then wrote her story and revised it with the help of her teacher. Her favorite part was being able to write about animals. She hopes readers will think it's funny and will learn that sheep are useful. She learned that writing is fun and that there are more than a billion sheep in the world. Ellie is most looking forward to going to the Capitol and signing autographs. She's looking forward to meeting our legislators and seeing her story's illustrations.

Hannah Butler, Nadia Spencer, Sarah Baron

Inderkum High School
Rachel Rodriguez, Art Teacher

The Mystery of Merino's Wool was illustrated by three students in Inderkum High School's art department. Hannah was the lead on the project and Nadia and Sarah supported her with the illustrations. Hannah started by researching sheep breeds so that she would have accurate representations for the illustrations. She then consulted with the other students and her teacher to prepare the sketches. They used watercolor for the illustrations. The students really enjoyed the challenge of this specific project.

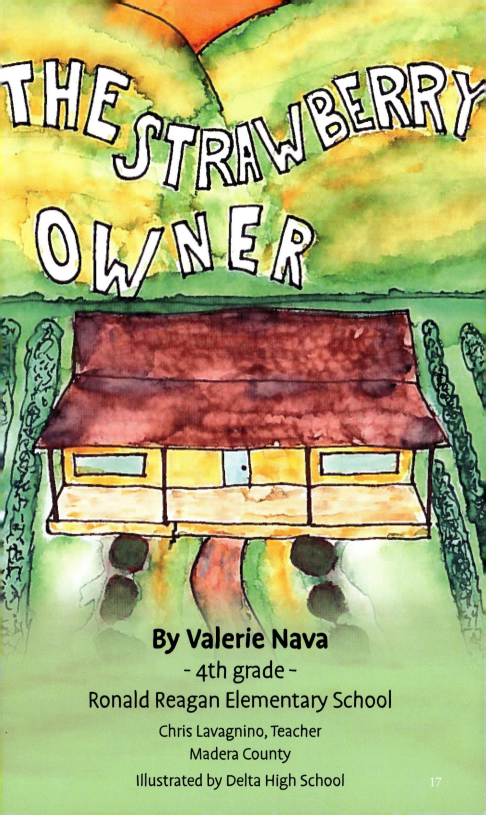

THE STRAWBERRY OWNER

By Valerie Nava
- 4th grade -
Ronald Reagan Elementary School
Chris Lavagnino, Teacher
Madera County
Illustrated by Delta High School

17

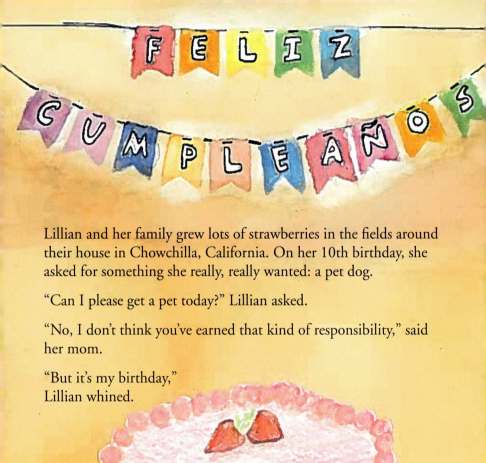

Lillian and her family grew lots of strawberries in the fields around their house in Chowchilla, California. On her 10th birthday, she asked for something she really, really wanted: a pet dog.

"Can I please get a pet today?" Lillian asked.

"No, I don't think you've earned that kind of responsibility," said her mom.

"But it's my birthday," Lillian whined.

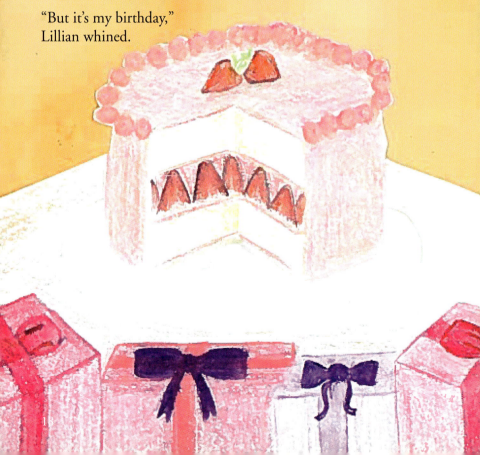

"Since you don't do your chores, you have not shown how responsible you can be," said her dad.

"Remember, strawberries are our livelihood. We need to have our patches ready for harvest each day during the springtime in the Central Valley."

"But Dad, I learned at school that most California strawberries are grown in coastal regions, like Monterey and other coastal counties. Why is our patch so important?"

"Mija, Central Valley strawberries are just as important – and delicious – to the people who live here," explained Dad.

Lillian barely heard his answer, because she was already starting her chores.

One Saturday morning, one of her chores was to pick strawberries by hand. As she picked them one by one, she knew that she needed to put the ripe ones in her basket immediately; otherwise she might be tempted to eat them all! As she worked, she heard a small sound from the ground.

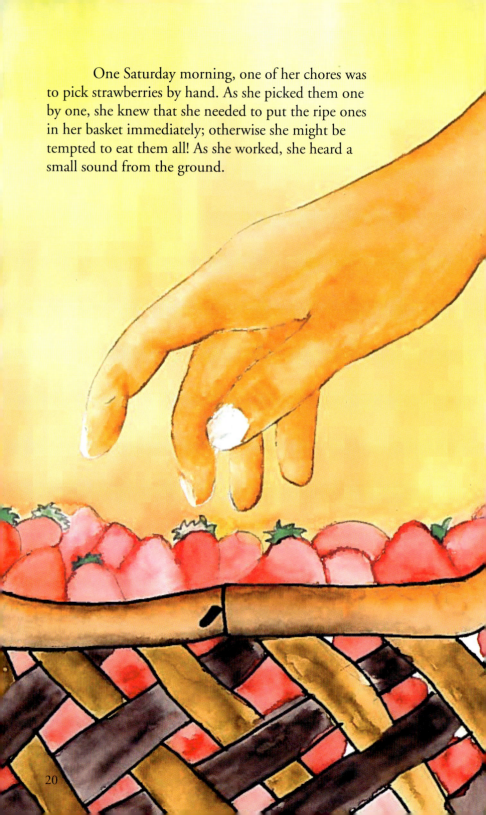

"Hi," said a sweet voice.

"Who said that?" questioned Lillian.

"Mmmm me," was the whispered response.

"Whooo?" questioned Lillian. "Where are you?"

"I'm down here," whispered the strawberry.

Lillian looked down and saw a strawberry rolling around in the small patch by her foot.

"Am I seeing things?" said Lillian.

"Nope," said the strawberry.

"Okay, I think I am going to faint," explained Lillian.

"Please don't fall on me," said the strawberry. "I don't want to become jam!"

After a moment, the strawberry asked, "I was just wondering: What is that stuff on your face? It looks like you got rained on, but the sun is shining!"

"I've been crying. But maybe you can help me! I really want to get a pet, but my parents won't let me," said Lillian.

"Why won't they? You seem berry responsible to me. I watched how careful you were at picking my cousins and how gently you placed them in your basket," said the strawberry.

"They said that I am not responsible and that I do not do my chores," said Lillian.

"Well, is that true?" asked the inquisitive berry.

"Well, I guess I have been kind of lazy lately," considered the girl. "But it is important for everyone to do their part on the family farm. I sure don't want my mama and papa having to do extra work because of me." Lillian thought about all the long hours her parents worked on the farm each day.

"Yeah, I really do need to start helping more."

"That is a great idea! Then they will see how responsible you really are," said the strawberry excitedly.

Lillian began to brainstorm all the extra things she could do around the property.

23

During the morning, she and the berry talked about the history of strawberries and how they got their names. As she was packing things away for the day, she had one final question for her little red friend.

"I've always wondered: How many seeds do you have?"

"Well, I'm not sure. I was told by my great-grandberry that we sometimes have as many as 200 seeds."

"Wow, imagine how many new berries might grow from you!"

The strawberry blushed.

For the next several weeks, Lillian did all of her chores and a few extra things to help out her parents around the farm. Her parents noticed.

"We are very proud of you, Lillian. You have been much better with your chores," said her dad.

"Oh my – you did all your work. I think you deserve a pet," said Lillian's mom.

"Thank you so much! I really appreciate it," said Lillian in excitement.

And later that day, Lillian got to go to the pet store to choose her puppy.

Now each day when her chores are done, the three friends take afternoon walks together. Lillian's parents often look out from the house admiring the change in their daughter's attitude and of her maturing attitude about daily living on a farm. But no matter how much they try, her parents can still not figure out what the little red bump on the dog's back really is.

About the Author:

Valerie Nava, age 9

Fourth grade student Valerie Nava got the idea for her story, *The Strawberry's Owner*, from her love of strawberries and her wish for a puppy. Much like the character in her story, Valerie has always wanted a pet puppy and her parents have said that caring for a puppy is a big responsibility. The main character was named after one of her friends, Lillian. Her favorite part of writing her story was when Lillian's parents saw how she had matured, and she was able to get a puppy! Valerie learned so many facts about strawberries during her research for the story. For instance, she learned that strawberries have an average of 200 seeds per berry. She also talked with her teacher and learned about some of the differences between strawberries grown on the coast and those grown in the central valley.

Valerie has always wanted to be a published author. She made her own book when she was in second grade and is very much looking forward to becoming a published author through the *Imagine this…* Story Writing Contest.

About the Illustrators:

James Clark, Joy Taylor, Elizabeth Mandujano, Zoey Mills

Delta High School | Corrie Soderlund, Art Teacher

The *Strawberry's Owner* was illustrated by four talented artists at Delta High School. James, Joy, Elizabeth, and Zoey started by reading through the story and brainstorming composition and layout ideas. The students divided the work by discussing which aspects of the story were intriguing to each of the students. For example, one student likes creating landscapes while another student prefers to draw objects that are close up. The students divided the work based on their strengths and continued to work together and check in with each other throughout the process. They really enjoyed working together and collaborating on the project. The students used pencil, watercolor, and ink for the illustrations. They thought the story was cute and that the author is very imaginative.

-ONE-
Amazing Family

By Olivia Piazza
- 5th grade -
Gratton Elementary School

Sheila Amaral, Teacher
Stanislaus County
Illustrated by Woodland High School

One night after dinner, my sister, Izzy, and I were sitting on the couch with my family. Dad and Grandpa were talking about our family history. They were looking at some old pictures Grandpa found while cleaning his office.

"This one is of my dad, whose parents immigrated to America from Sicily, Italy. They landed in Portland, Oregon. They came here in the 1900s. This photograph is very old," said my grandpa.

1947,

"Oh, that looks like you, Dad," I said as I leaned over. It was an old black-and-white photo. On the back it said 194 with a stain covering the rest of the year and what looked like a signature. "What does that say?"

"It says 1947, Tony Piazza," Grandpa explained. "My dad's family moved from Portland to a farm in Mountain View.

33

Then in the 1950s, we moved here to the small town of Denair. That's when we started this farm."

"How did the farm start?" I wondered.

"My parents had a farm by the river. It was the second time in three years that the farm had flooded. So, in 1950 they found this place and moved here," stated Grandpa.

I replied, "That's so cool!" I was learning so much by listening to my grandparents and dad talk.

Grandpa was saying, "California is the best place for almonds to grow because of our climate. We have hot summers and cool winters that almonds need."

Eventually, Dad said, "Well, girls, it's time to go." I gave Grandma and Grandpa a hug goodbye.

The next morning, I went to school. When I got home, I didn't hear the loud noises of machinery. It wasn't until dinner that I knew what happened.

"The huller broke down," Dad explained. "Nothing to worry about. We'll have it fixed by tomorrow."

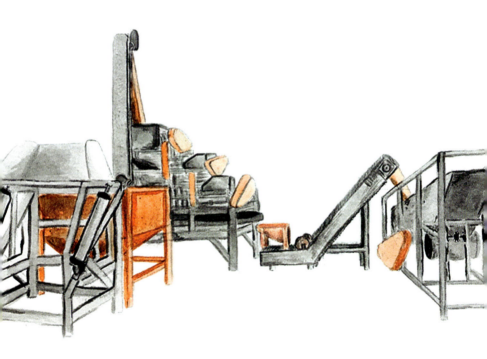

The next day, the huller still wasn't running. I thought by the time I got home, it would surely be back on. After school, the huller was quiet as a ghost town. I did my homework and read in my room. Outside, I saw Dad with a worried look on his face. He was talking to a worker. Suddenly, he started walking toward the house.

Three days later, Dad sadly said, "I don't know how to fix the huller." I was thinking about the picture of Grandpa Tony. "Grandpa!" I announced loudly.

"What?" asked Izzy.

"We could ask Grandpa how to fix the huller!" I yelled.

"Calm down, Olivia," Mom said.

"No, that's a great idea!" exclaimed Dad.

Right away, my family went next door to my grandparents' house.
I was the first there. I opened the door; Dad was right on my heels.

"Grandpa!" I shouted.

"Mom! Dad!" called out my dad.

"We're right here," Grandma said.

"What's going on?" Grandpa asked.

"The huller broke down and we don't know how to fix it! Do you know how to make the repairs?" questioned Izzy.

"What's wrong?" asked Grandpa. Dad told him, "It just stopped. Completely stopped."

"Oh no," he whispered.

"What?" I whispered back.

"This has happened only twice before, and it was long ago," he said quietly.

"But you know what to do, right?" Izzy asked.

"Right?"

"No," he muttered. "No, I don't know what to do." We sat in silence, all thinking.

Finally, it dawned on me. "What about, what if, somehow…"

"What!" gasped everyone.

"Maybe we could search the office for ideas," I finished.
Grandpa said, "This photo was in a file with other documents."

"Let's look for it!" cried Grandma.

So, after some intense searching and scouring the office for the file,
Mom held up a thin file.

"That's it!" Grandpa announced.
We all started reading the file.
I saw it said, To fix the broken
huller, check a secluded conduit.
There might be damage such as…

"Mice!" announced Grandpa.
"I know how to fix this!"
The next day, Dad and Grandpa fixed
the huller. It was loud and annoying,
but I couldn't be happier to hear it.

Later that day, Grandpa and I added two pictures to a brand-new file with a note. One was the picture of Grandpa Tony. The other was a picture I had taken on my sister's camera. And the note, in my own handwriting, said, If the huller breaks down, all you need is an amazing family to help you fix it.

About the Author:

Olivia Piazza, age 11

Olivia's story, *One Amazing Family*, was inspired by her own family and their almond huller. Her family lives on an almond ranch and so Olivia was able to learn a lot about almonds and the hulling and shelling process from her family. She started preparing for her story by researching facts about almonds and almond hullers and talking with her grandparents. Then, she started writing and editing!

Olivia's favorite part of writing her story was asking questions about her family's past. She was able to learn about her grandpa's life and is excited to share this history with readers. She hopes that readers enjoy learning about her family, and she hopes that they learn about the growing and shelling process. Olivia is looking forward to having her story published!

About the Illustrators:

Addie Ferrer & Lauren Peña Turner

Woodland High School
Scott Coppenger, Art Teacher

Addie and Lauren are both seniors in the AP Art classes at Woodland High School. In illustrating *One Amazing Family*, they learned about the almond hulling process in order to accurately draw the almond huller. Addie created the backgrounds and inanimate objects of the story while Lauren was in charge of illustrating all the people. They took inspiration from some of their favorite children's books and used watercolors and Sharpies to create the artwork. Lauren and Addie were very excited to bring the author's story to life. They both felt honored to be asked to illustrate this story.

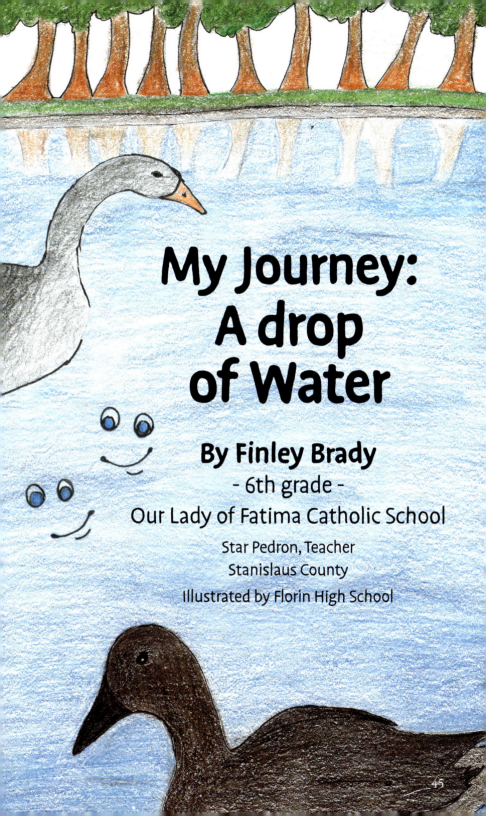

My Journey: A drop of Water

By Finley Brady
- 6th grade -
Our Lady of Fatima Catholic School

Star Pedron, Teacher
Stanislaus County

Illustrated by Florin High School

Hi, my name is Sam. Let me tell you a little about myself. I am a drop of water living in the Pacific Ocean. A long time ago, I used to be a tiny snowflake that fell from a cloud. Then, I left my family and came to rest on Mount Shasta. It was a dark night when I found myself on Mount Shasta without my friends or family.

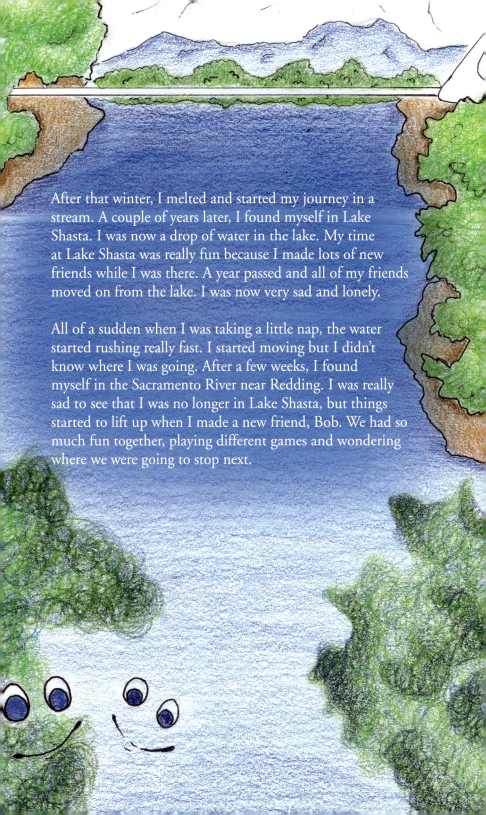

After that winter, I melted and started my journey in a stream. A couple of years later, I found myself in Lake Shasta. I was now a drop of water in the lake. My time at Lake Shasta was really fun because I made lots of new friends while I was there. A year passed and all of my friends moved on from the lake. I was now very sad and lonely.

All of a sudden when I was taking a little nap, the water started rushing really fast. I started moving but I didn't know where I was going. After a few weeks, I found myself in the Sacramento River near Redding. I was really sad to see that I was no longer in Lake Shasta, but things started to lift up when I made a new friend, Bob. We had so much fun together, playing different games and wondering where we were going to stop next.

47

Then one day, a truck came along and sucked Bob and I out of the river and pumped us into a water tower. I was very curious if I was ever going to get out of the water tower in Redding and so was Bob. A whole year passed while Bob and I hung out in the water tower. It was dark in there and we wondered if we would ever see the light of day again.

Suddenly, one day I woke up and I was in a water truck with Bob heading south on I-5. We were on our way to Williams. We were both wondering what Williams was going to look like when we got there. Williams ended up being nice and a very sweet town. There were also lots of crops to water in Williams, like rice and tomatoes.

We ended up being sprayed onto a road and trickled into a rice field. We hung out in that rice field and a bunch of different irrigation canals for another couple of years. I had the best time in Williams with Bob.

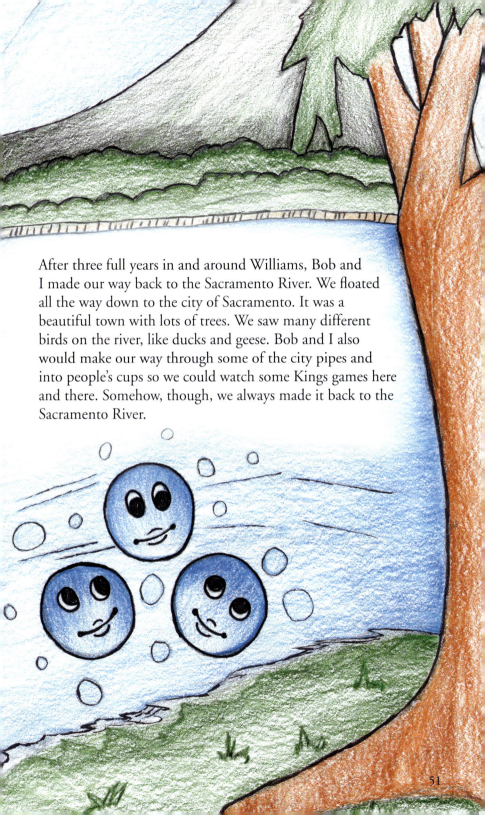

After three full years in and around Williams, Bob and
I made our way back to the Sacramento River. We floated
all the way down to the city of Sacramento. It was a
beautiful town with lots of trees. We saw many different
birds on the river, like ducks and geese. Bob and I also
would make our way through some of the city pipes and
into people's cups so we could watch some Kings games here
and there. Somehow, though, we always made it back to the
Sacramento River.

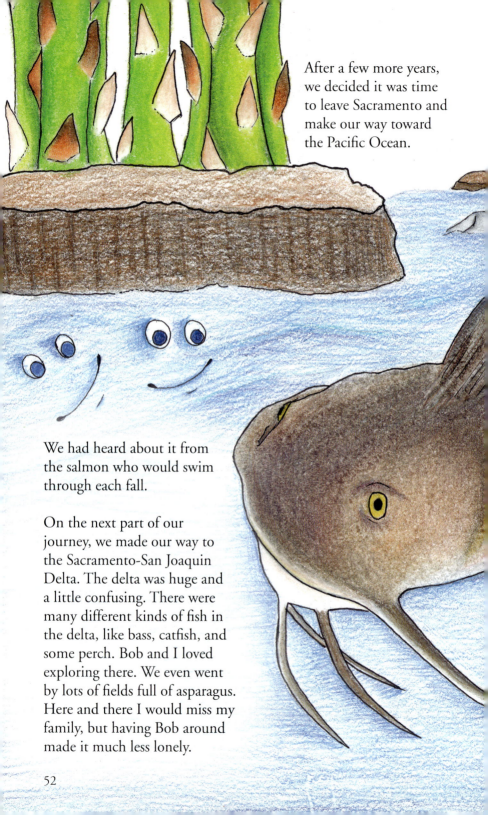

After a few more years, we decided it was time to leave Sacramento and make our way toward the Pacific Ocean.

We had heard about it from the salmon who would swim through each fall.

On the next part of our journey, we made our way to the Sacramento-San Joaquin Delta. The delta was huge and a little confusing. There were many different kinds of fish in the delta, like bass, catfish, and some perch. Bob and I loved exploring there. We even went by lots of fields full of asparagus. Here and there I would miss my family, but having Bob around made it much less lonely.

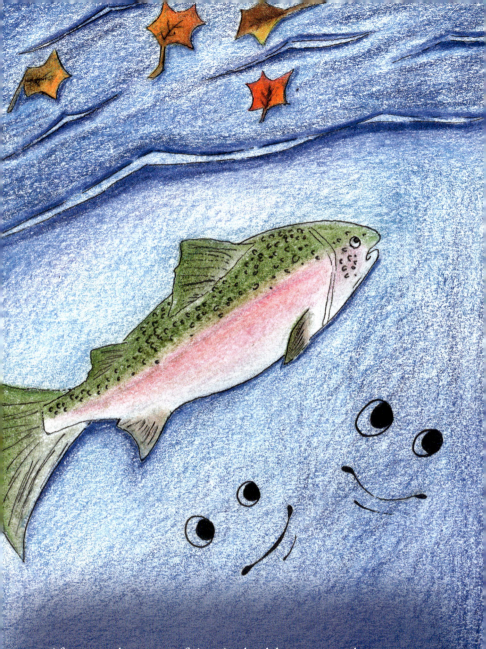

After a good amount of time in the delta, we moved to our final destination, the Pacific Ocean. We dropped in right at San Francisco. I was so happy to be there with my best friend, Bob. He and I saw the Golden Gate Bridge and lots of salmon on their way to the Sacramento River.

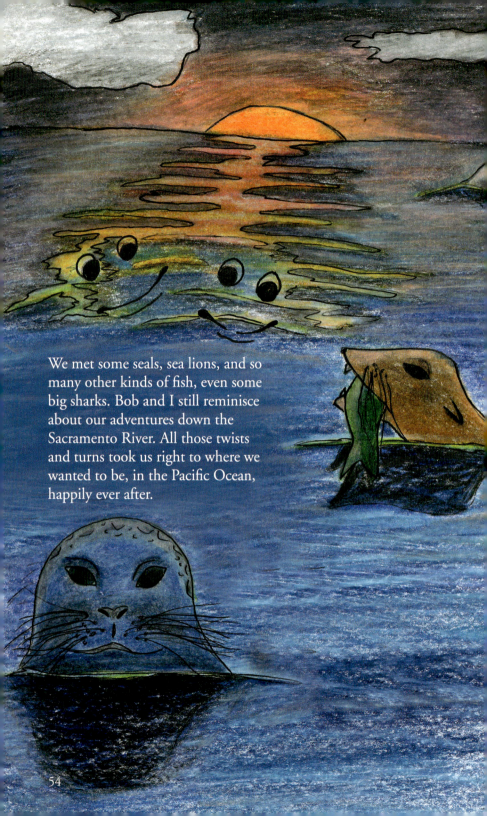

We met some seals, sea lions, and so many other kinds of fish, even some big sharks. Bob and I still reminisce about our adventures down the Sacramento River. All those twists and turns took us right to where we wanted to be, in the Pacific Ocean, happily ever after.

About the Author:

Finley Brady, age 12

Sixth grade student, Finely Brady got the idea for his story, *My Journey: A Drop of Water*, from all the places in Northern California that he has visited with his family. He started his writing process by outlining his idea and formatting his ideas into a story. His favorite part about the writing process was the places that he takes the reader in the story because it brought back so many memories from trips with his family and friends.

Finley hopes that readers will learn that friendship is really important in life and that water is an important part of California agriculture. From his research, he really enjoyed learning about the role and importance of water in agriculture and about the Sacramento River.

About the Illustrators:

Emilie Ly, Pa Zong Vang, Zolpenoor Shafaq

Florin High School | Alexandra Pease, Art Teacher

Before reading the story, *My Journey: A Drop of Water*, the students at Franklin High School knew a little bit about water but were not expecting to illustrate a snowflake or water tower. They really enjoyed the story. Through the process of creating the illustrations, Emilie, Pa, and Zolpenoor were able to learn more about the Sacramento and San Joaquin River Delta systems along with many locations like Mount Shasta. To begin, the students looked at the story and created an illustration plan based on the paragraphs. They agreed to divide the work based on their individual strengths. Emilie drew the scenes and Pa and Zolpenoor added the color. The students loved the story and all the details the author included. Overall, they really enjoyed this experience and look forward to sharing their illustrations with the author.

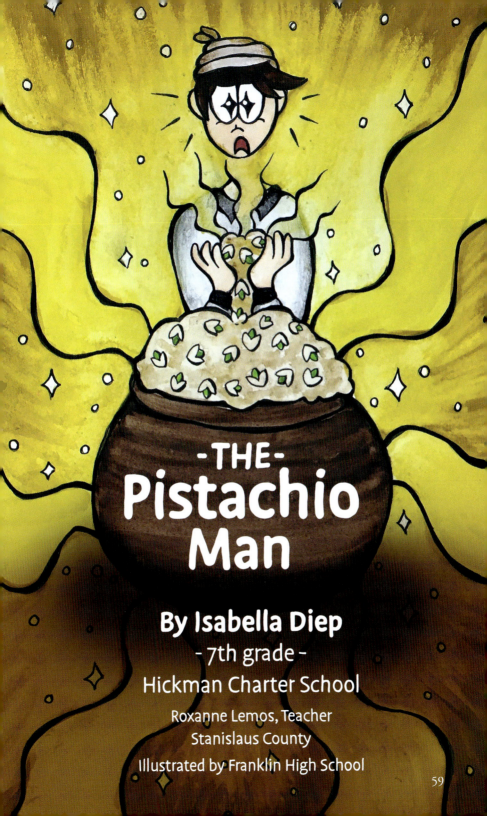

-THE-
Pistachio Man

By Isabella Diep
- 7th grade -
Hickman Charter School

Roxanne Lemos, Teacher
Stanislaus County

Illustrated by Franklin High School

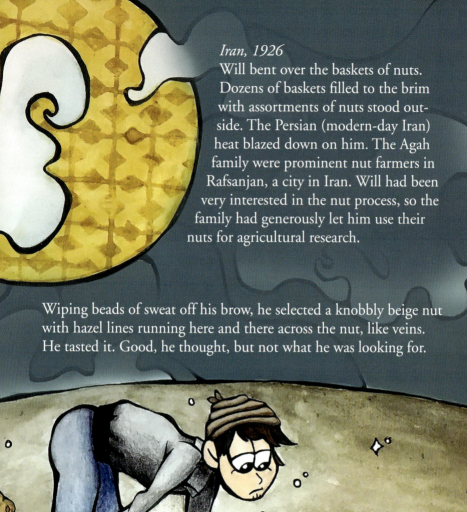

Iran, 1926
Will bent over the baskets of nuts.
Dozens of baskets filled to the brim
with assortments of nuts stood outside. The Persian (modern-day Iran)
heat blazed down on him. The Agah
family were prominent nut farmers in
Rafsanjan, a city in Iran. Will had been
very interested in the nut process, so the
family had generously let him use their
nuts for agricultural research.

Wiping beads of sweat off his brow, he selected a knobbly beige nut
with hazel lines running here and there across the nut, like veins.
He tasted it. Good, he thought, but not what he was looking for.

Cautiously he
examined the nuts.
As he picked through
the baskets of nuts, he
tasted hazelnut, almond
and peanut. Those nuts
he was quite familiar with,
though they tasted better
than the ones in Maryland.

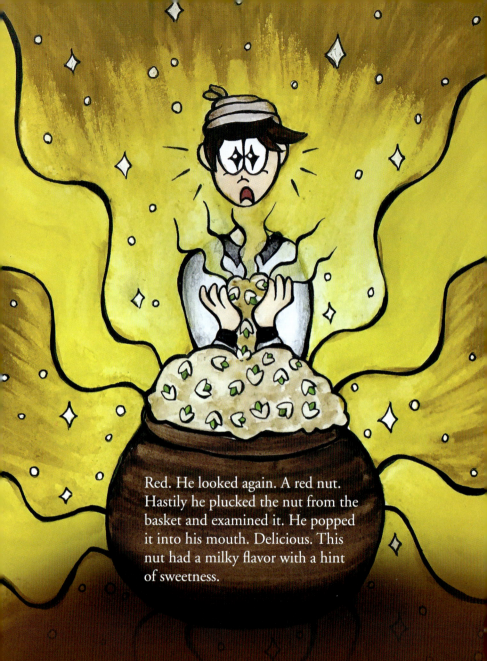

Red. He looked again. A red nut. Hastily he plucked the nut from the basket and examined it. He popped it into his mouth. Delicious. This nut had a milky flavor with a hint of sweetness.

Greedily he hunted the other baskets for the red nut, cramming them into his mouth. Satisfied, he summoned Parvaneh, the 14-year-old daughter of Shahryar Agah, the owner of the nut farm. Parvaneh had been entrusted with the task of answering all Will's questions and showing him around the farm.

Parvaneh appeared a few moments later with her younger brother, 4-year-old Farrokh, in her arms. "Parvaneh, what is this nut?" William asked. She answered in her heavily accented English,

"It is the pistachio nut."

"The pistachio nut is new to me. Would your family allow me to take some home with me?"

She tilted her head, while little, tired Farrokh fussed.

"I will go ask my father," she responded at last.

When Shahryar Agah arrived, Will expressed his desire to take the pistachio back home. Shahryar and William discussed it deep into the night.

The next morning, the decision was made. William would be allowed to take the pistachio nut back to America. He would take some pistachio seeds, and hopefully be able to plant them in the American soil.

California, 1926

Lawrence eagerly ripped open the envelope. He was sitting on his kitchen stool, a pile of mail in front of him. Sorting through the mail, he came across an envelope from William.

William Whitehouse and Lawrence Wilson were best friends. They trusted each other, and told each other everything. Lawrence had been a bit hurt when Will had left without a trace, but here was something from Will. A letter tumbled out of the unsealed envelope. As fast as an arrow whizzing through the air, Lawrence speedily snatched up the letter. Unfolding it he read,

1926 Dear Lawrence,
My most cherished friend, I sincerely apologize for not telling you about my trip to Persia. I went because I'd heard about their famous nut farms, and as you know, since I was studying agriculture,

65

I thought it would be fun to take a trip to the Middle East. I hoped to learn about the subject more. I found a unique nut. The nut growers call it the pistachio nut. I've never tasted it before, so I have decided to take some seeds back to California and grow them. I hope you approve of my plan because you are the first one I've told beside the nut farmers who gave me the permission. See you soon.

Yours truly,
William E. Whitehouse

Lawrence read the letter again. And again. And again. He should have known good ole' Will was up to something. But never in his life would he think Will was going to travel to the Middle East. He knew one thing, though: Mrs. Whitehouse, Will's mother, deserved to know. He left the house.

California, 1929

News of Will's return spread like wildfire. Everywhere in his hometown, people wanted some of the pistachio seeds. They planted them. They watered them. The seeds grew. In people's yards and farms, many pistachio trees were planted. People loved the pistachios and harvested many.

Factories opened, businesses started, people farmed and planted. Many years later, California was one of the main pistachio-producing states in the United States.

Still, even though he didn't know it now, pistachios would become a very popular snack. People would buy them everywhere, eat them everywhere, and sell them everywhere.

Will was proud, prouder than he had ever been in his whole life. He was the "Pistachio Man."

About the Author:

Isabella Diep, age 12

Isabella's story was inspired by her love of pistachios! Through her research, she learned about the man who brought pistachios to America (The Pistachio Man) and the idea for her story was born. She started her writing process by free-writing based on her idea. Then, she started adding a plot, characters, details and dialogue. Her favorite part of writing the story was the character profiles. She really enjoys brainstorming names, ages and all of the details surrounding the characters.

Isabella hopes that readers will learn how pistachios came to California and that many different commodities have come to California from other countries through a similar process. Isabella has always really enjoyed writing and is really looking forward to being a published author and seeing her story in print!

About the Illustrators:

Liv Bryant, Connie Weng, Samantha Jang

Franklin High School | Derek Bills, Art Teacher

Before illustrating *The Pistachio Man*, Franklin High School Art Department students Liv, Connie and Samantha knew very little about the history of pistachios. Through the story, they learned the origin of pistachios and how they became a popular snack in the United States. All three students started by brainstorming ideas for the illustrations and then Liv, the head illustrator, started producing the layout of each piece. Samantha then inked the drawings and Connie added the color to finish them. The students used watercolor, ink and colored pencil for the illustrations. They really enjoyed seeing the story come to life and working together as a creative team.

Laura & Lana:
California Fairies
of Flora & Fauna

By Nathan Tanega
- 8th grade -

Our Lady of Fatima Catholic School

Star Pedron, Teacher
Stanislaus County

Illustrated by Sheldon High School

There once were two fairies named Laura and Lana
Their hope was to help our state's flora and fauna
Nurturing plants and helping animals: their power
California their kingdom, their home in Bellflower.

Lana's focus was animals and all things therewith
With her magic touch, cattle and calves are ranked fifth
And also in things like sheep, turkeys, and hogs,
And horses and chicken, but sorry, no frogs!

75

And then there's the egg and here's a quick guide
You can eat it poached or hard boiled or scrambled or fried
An omelet or eggs Benedict, there are so many ways
I know I, personally, could eat them for days!
And when you drink milk
(and I think that you should)
It is not only delicious, but it does a body good!

In addition to beef, we produce poultry and dairy,
All this, and much more, is made with the help of our fairy.

Laura, the other fairy, uses charms on our plants
And that might make you think of a fern at first glance,
But you would be wrong because this group is quite vast
Grapes and melons are a part of this cast.
Of course, there is corn and almonds and grains
And soybeans and vegetables and other food for our brains

Oranges, strawberries, apricots, lettuce
Figs, prunes, pistachios, and even asparagus
From avocados to lemons to peaches galore
These are just some of the things, and there's more!

In addition, there are carrots, which are good for the eyes
And cotton, which turns into socks, tees, and ties

With grapes, of course, comes all varietals of wine
And garlic, which has always been a favorite of mine!
Plums and tomatoes and broccoli and beets
And would you believe this list is not even complete?

Floriculture is also a big focus to Flora
This includes plants and flowers (which I just adore-ahh)
From poinsettias to orchids to a rose and a lily
An azalea, but not Iggy, because that would be silly!

Ranunculus, sunflowers, and carnations and more
And dahlias and anemones add beauty to any décor
And with this crop, we are number one in the nation
And they are wonderful to give on any occasion!

Thanks to the fairies, our state's export value is in billions
And we are the top ag-producing state (so
shouldn't that be zillions?)

And while these fairies work hard and are charmers
It isn't elves or gnomes, but people, especially the farmers
That work hardest of all with their own magical wands
That would be tractors and planters that run dusk 'til dawn.

From grain augers and harrows to balers and more
Their powers seem enchanting like tales in a lore
But with all they produce, it really is no wonder
California ag is number one and that is no blunder!

So, alas, it is the hard work that rules the day
No one would argue that, wouldn't you say?

So let us say thanks to both Laura and Lana
With their support, we are tops in both flora and fauna!

About the Author:

Nathan Tanega, age 14

Nathan's poem, *Laura and Lana: California Fairies of Flora and Fauna*, was inspired by his mom. His mom uses the term 'Flora and Fauna' and so he thought it would be interesting to write a poem based on the phrase. He started by researching California agriculture and wrote down as many facts as he could about California agriculture—ranging from plants, flowers and animals. After that, Nathan started drafting the poem and finding words and facts that rhymed. His favorite part of writing the poem was putting the rhyming words together.

Throughout his writing process, Nathan learned so many fun details about California agriculture and hopes that readers will learn as many fun facts about agriculture as he did while writing. Nathan had no idea that California produced so many kinds of flowers and that California is the number one agricultural state in the nation. He is really looking forward to seeing his poem illustrated!

About the Illustrators:

Sophia Reynoso-Lopez and Akilah York
Sheldon High School | Kelsey Dillard, Art Teacher

Before illustrating the poem, *Laura and Lana: California Fairies of Flora and Fauna*, Akilah and Sophia did not know what to expect from the poem. Both students are Senior AP Art students at Sheldon High School. To Illustrate the poem Akilah and Sophia first divided the story into sections. Then, they both did research to accurately illustrate the poem. After several sketches they began to watercolor the drawings. Both Sophia and Akilah learned that California is one of the main contributors and producers of many commodities. Sophia enjoyed drawing something out of her comfort zone, while Akilah loved the experience because she wants to illustrate children's books as a career. Overall, illustrating the book was a positive and challenging experience for Sophia and Akilah.

86

Alissa Alfalfa's Big Journey

Honorable Mention

By Ellie Gomes

8th grade – Scott Valley Junior High

Amy Hurlimann, Teacher

On a warm spring morning in the middle of May, there was a field of alfalfa sprouts on a farm in Scott Valley. All of the sprouts were happy to be out of the ground and looking around, all except one. Her name was Alissa Alfalfa and she did not feel special. Alissa wanted to be unique; she did not like that there was a whole field of alfalfa that looked just like her. A bird that often visited happened to notice the sad little alfalfa plant.

"Hi, my name is Maggie, the magpie. I come to visit this alfalfa field often, and I can't help but wonder why you are sad," inquired the bird.

"I am sad because I want to be unique. I don't want to look like everyone else. I feel like no one wants me because there are so many like me," Alissa explained sadly.

"Are you kidding? You are California-grown alfalfa!" Maggie exclaimed.

Alissa looked up at the bird. "What do you mean?" she asked curiously.

"Do you not know how important you are not only to the U.S. but the whole world? You are what feeds the world, and you are extra special because you are a very beautiful kind of alfalfa that only grows in California!" the bird explained.

"Wow! I really am important!" Alissa happily replied.
"Soon enough you will be cut, dried and grouped together with

all your friends, then shipped on trucks and boats to see the world! The hay you make is so special that it is wanted all over this country and many others," Maggie told Alissa.

This made Alissa really happy; she must be unique after all. Now she was so proud to be an alfalfa plant and wanted to share what she had learned from the magpie with all her friends.

"Everyone, guess what?" Alissa asked excitedly.

"What?" asked the alfalfa sprouts.

"We are going to travel! My friend Maggie says that California alfalfa is wanted all over the world!" she beamed.

All the other young alfalfa plants looked around at each other in awe. Many said things like, "Wow!" but most said nothing at all because they were speechless. The young plants were all amazed.

After about 50 days of growing, the alfalfa sprouts became big, beautiful plants that were ready to be cut, dried, and baled. All the alfalfa plants were excited and a little nervous to see where they were going to be sent. Their friend Maggie came to visit often and was there all day when they were being cut. When all were on the ground, they were raked and left out in the sun to dry. Then one day, a loud baler came and grouped them together, making bales. The bales were loaded on a truck that drove many hours down to the Bay Area, occasionally stopping to deliver alfalfa bales. The whole journey, Maggie stayed by Alissa's side.

"Wow! What is that, Maggie?" all the alfalfa plants would ask when dazzled by things they had never seen before.

Usually Maggie would say things like, "That is a town where people live" or "That is a place where the trucks fill up the diesel tank, so they can keep driving us." When the truck full of beautiful Scott Valley hay finally arrived at the Port of Long Beach, it was sorted and loaded onto ships that would cross the Pacific Ocean. They had never seen such tall buildings! Maggie told Alissa the ships would travel to Japan, China, Arabia, Korea, and other countries needing them.

"Wow!" they exclaimed.

"Now our ship will be going to Japan!" Maggie told them.

After they said goodbye to their friends, the ship started its journey to Japan. Along the way, Alissa was amazed by the many ocean animals Maggie pointed out. After 30 days, they docked in the Port of Tokyo and were loaded into trucks once again. Alissa watched as some of her friends were unloaded on a dairy farm, others on a beef ranch, and a few at a feed store. Finally, Alissa and Maggie arrived in a fancy stable yard. Maggie took to the sky to look around.

"Oh my goodness! Alissa, you won't believe it! We are at the Tokyo Racecourse. The Japan Cup is next week, and you have been sent to help the beautiful horses prepare for the race." Maggie sang with delight.

Alissa beamed with pride. Never in her life had she been so happy.

Author
Ellie Gomes (8th Grade)

Illustrator
Ovava Tonga

GLOSSARY

Agriculture: the science or practice of producing resources including the five F's: Food, Fiber, Flowers, Forests, and Fuel.

Bale: a large bundle prepared for transportation and storage.

Canal: an artificial waterway used to bring water to a field to irrigate land.

Climate: the general atmospheric conditions for a location, including rainfall, temperature, humidity, etc.

Crochet: the process of creating textiles by using a crochet hook and yarn or strands of other materials.

Crop: plants that are grown and harvested by farmers.

Fern: a seedless, flowerless vascular plant.

Floriculture: the growing of flowers.

Grain Auger: a long tube with a spiral shaft down the middle used to raise and transport grain from the ground to the top of a grain bin.

Grain Harrows: a machine pulled by a tractor that prepares a seedbed for planting.

Harvest: the process of gathering mature crops from the field.

Huller: a large machine that removes the hulls and shells from nuts.

Immigrate: the act of coming to live permanently in a foreign country.

Irrigation: the act of adding water to crops to help with growth.

Livelihood: the means of securing the necessities of life.

Machinery: a group of large machines or parts of a machine that make it work.

Port: a town or city with a harbor where ships load and unload products.

Secluded Conduit: tubing used to protect electrical wires.

Shearing: the process by which the woolen fleece of a sheep is cut off.

Water Tower: an elevated tank used for water storage and delivery of water through a piped system.

Wool: the hair of animals such as sheep and goats.

ACKNOWLEDGMENTS

California Foundation for Agriculture in the Classroom (CFAITC) would like to acknowledge the many people who contributed to the success of the 2019-2020 *Imagine this...* Story Writing Contest and the *Imagine this... Stories Inspired by Agriculture* book.

——— MANY THANKS TO: ———

Imagine this... Regional Coordinators

Jacki Zediker, Region 1   Sandra Gist-Langiano, Region 3
Doni Rosasco, Region 2   Mary Landau, Region 4

High School Art Programs

Delta High School, Clarksburg
Franklin High School, Elk Grove
Florin High School, Sacramento
Woodland High School, Woodland
Sheldon High School, Elk Grove
Inderkum High School, Sacramento

CFAITC Staff

Judy Culbertson, Publisher, *Imagine this...* book
Mindy DeRohan, Coordinator and Editor,
Imagine this... Story Writing Contest and book

Design & Layout

Lisa Daniels Design

CFAITC Board of Directors

Jamie Johansson
CFAITC President
Lodestar Farms

Correen Davis
The Gorrill Ranch

Mark Dawson
*California
Farm Bureau
Federation*

Martha Deichler
Retired Superintendent
Borrego Springs USD

Shannon Douglass
Douglass Ranch

Bobbin Mulvaney
Mulvaney's B&L

Rick Phillips
Simplot

Jane Roberti
Roberti Ranch

Craig Thomson
*Zenith AgriBusiness
Solutions*

Kenny Watkins
Watkins Ranch

Becca Whitman
Raley's

ABOUT CALIFORNIA FOUNDATION FOR AGRICULTURE IN THE CLASSROOM

California Foundation for Agriculture in the Classroom (CFAITC) is dedicated to fostering a greater public knowledge of the agricultural industry. Since 1986, CFAITC has provided educators with free, quality teaching resources, along with professional development and grant opportunities. By equipping classroom teachers, after-school coordinators and other educators, CFAITC promotes student understanding of California agriculture.

CFAITC works with K-12 teachers and community leaders to help young people learn where their food comes from and how it arrives at grocery stores and restaurants. With this knowledge, students grow up with the ability to make informed choices. Through programs and resources, educators are encouraged to incorporate agriculture into lessons on various subjects and to explain the important role it plays in our economy and society.

Agriculture is an important industry in California. As more rural areas become urbanized, maintaining existing farmland and feeding the growing world population becomes a greater challenge. It is important to educate students about their environment and the opportunities agriculture provides. Students can explore agricultural careers based in science, business, engineering, communication and other disciplines.

WHY TEACH ABOUT AGRICULTURE?

Agriculture is everywhere. It is the food we eat, the clothes we wear, the houses we live in, and the plants we enjoy. A well-rounded education should include hands-on experiences and true-to-life learning at the most fundamental level. By exposing students to agriculture through engaging lessons and activities, we hope students will be inspired to continue to learn about food and fiber and, ultimately, gain an appreciation for all that agriculture provides.

Imagine this... Story Writing Contest
Annual Deadline: November 1

The *Imagine this...* Story Writing Contest aligns with the 4 C's of the Common Core State Standards—collaboration, creativity/innovation, critical thinking, and communication—while connecting students to the world around them! The contest meets the Common Core State Standards for narrative writing. Students (grades 3-8) write narratives to develop real or imagined experiences or events (*Common Core State Standards, W.3-8.3*) based on accurate information about California agriculture.

Every student who writes a story receives a packet of seeds and certificate of participation. Recognition is given to 48 students as regional winners and six are recognized as state winners. State-winning stories will be illustrated and published in the *Imagine this...* Stories Inspired by Agriculture book.

Entry Details:

- Stories must be written by one student author; no group entries allowed.
- Stories should not exceed 750 words.
- Stories should not be similar in theme to winning entries from previous years.
- Submit up to five stories from each classroom.
- Each story must have an entry form attached.

Winning stories must be:

- Titled and original, creative student work
- Fact or fiction
- Appropriate for classroom use
- Grammatically correct
- Related to California agriculture in a positive way
- Typed, preferably, or neatly handwritten
- Written without reference to registered trademarks

Please refer to CFAITC's website for past state-winning stories and a detailed list of the regional and state awards:

LearnAboutAg.org/imaginethis

ENTRY FORM

Send story and entry form to your Regional Coordinator (listed on back).

Student Name _____

Title of Story _____

Grade _____ Word Count _____

Teacher Name _____
(print first and last name)

Teacher Signature _____

School Name _____

School Address _____

City, State, Zip _____

School Phone   (   ) _____

County _____ Region Number (see map) _____

Teacher's E-mail _____

School District _____

How did you hear about this contest? _____

Hometown Newspaper _____

_____ Public School   _____ Private School   _____ Home School

Postmarked by November 1, annually

REGIONAL COORDINATORS

Send story and entry form, postmarked by November 1st, to your regional coordinator.

REGION 1
Jacki Zediker
7132 E. Louie Rd.
Montague, CA 96064
(530) 459-0529

Butte	Nevada
Colusa	Plumas
Del Norte	Shasta
Glenn	Sierra
Humboldt	Siskiyou
Lake	Sutter
Lassen	Tehama
Mendocino	Trinity
Modoc	Yuba

REGION 2
Doni Rosasco
16002 Hwy. 108
Jamestown, CA 95327
(209) 984-3539

Alameda	Napa	Santa Cruz
Alpine	Placer	Solano
Amador	Sacramento	Sonoma
Calaveras	San Francisco	Stanislaus
Contra Costa	San Joaquin	Tuolumne
El Dorado	San Mateo	Yolo
Marin	Santa Clara	

REGION 3
Sandi Gist-Langiano
P.O. Box 748
Visalia, CA 93279
(559) 732-8301

Fresno	Merced
Inyo	Mono
Kern	Monterey
Kings	San Benito
Madera	San Luis Obispo
Mariposa	Tulare

REGION 4
Mary Landau
330 East Las Flores Drive
Altadena, CA 91001
(626) 794-4025

Imperial	San Diego
Los Angeles	Santa Barbara
Orange	Ventura
Riverside	
San Bernardino	